MASE DISCOVERS EXOPLANETS

SEBASTIAN PEREZ

In Memory of Sixana Perez

On our last voyage together, Mase took us through the Solar System and beyond the Kepler Belt to learn about the closest Dwarf Planets to Earth. His spacecraft was ready to take us further away.

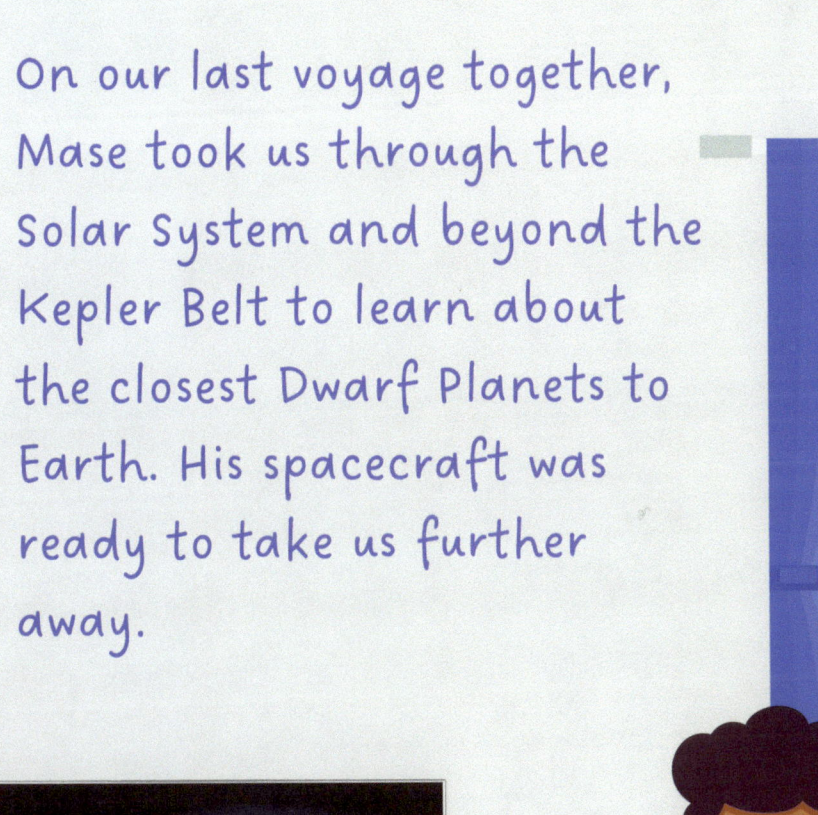

Mase was ready to speed into space and teach us everything he knew about Exoplanets. These types of planets are located beyond our Solar System. In order to get there Mase would have to travel light years beyond our system.

55 Cancri e

Our first Exoplanet lies 41 Light years from Earth. We arrive at 55 Cancri e, a world of lava with a sparkling sky. It completes an orbital period of 0.7 days around its star. It is located in the Constellation Cancer.

Gliese 1132b

The next Exoplanet on his list was in the Vela Constellation. Gliese 1132b is about 41 light years from Earth. Its orbital period was 1.6 days. Scientists believe it has a second atmosphere.

Upsilon Andromedae B

Our course has now turned 44 light years from Earth. In the Andromeda Constellation. We find ourselves gazing upon a "Gas Giant" dubbed the Land of Fire and Ice. The planet is in Tidal Lock with its sun, so one side is extremely hot, while the other side can range from minus 20 to 230 degrees.

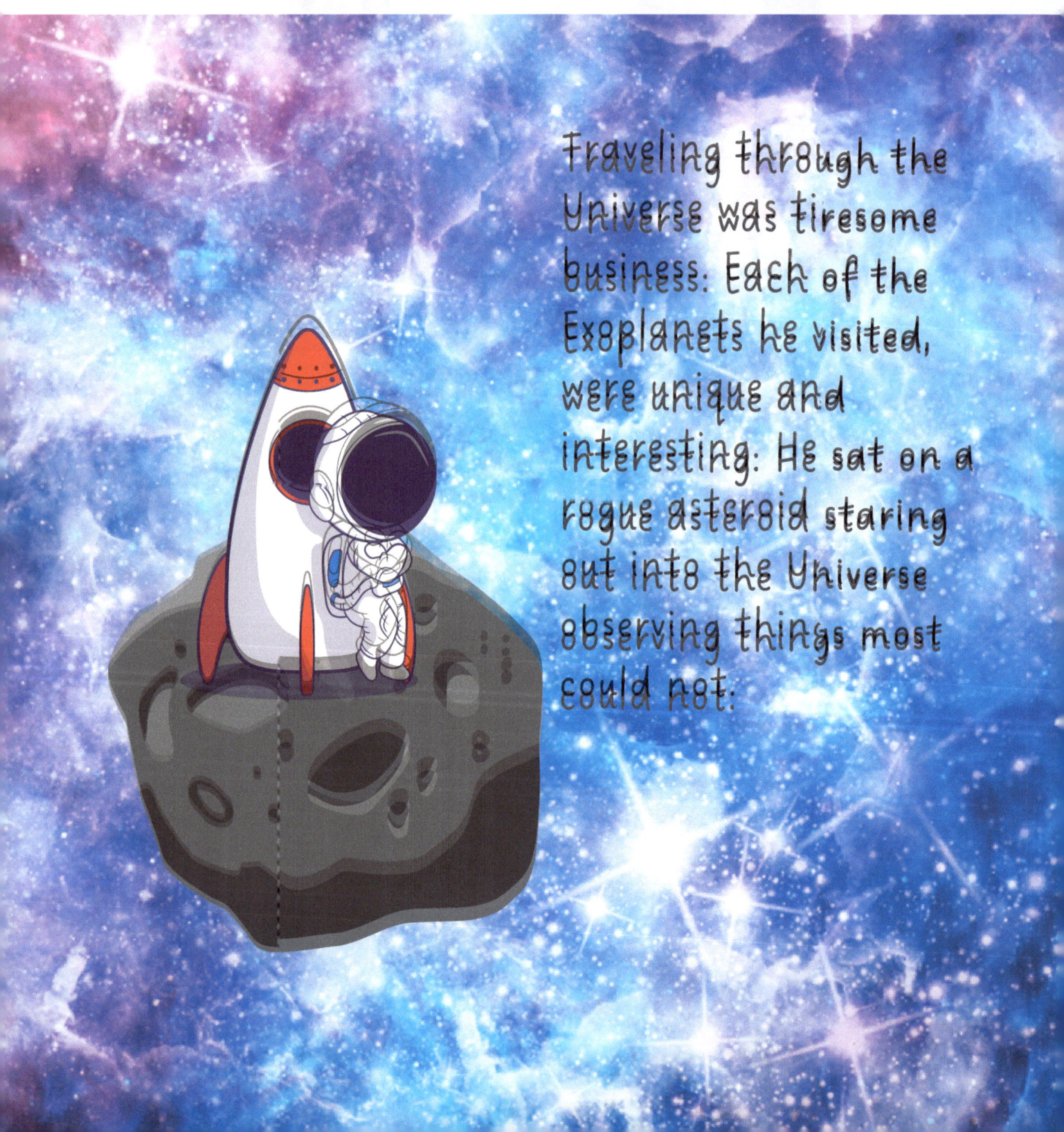

Traveling through the Universe was tiresome business. Each of the Exoplanets he visited, were unique and interesting. He sat on a rogue asteroid staring out into the Universe observing things most could not.

HD 189733B

Mase was ready to go even further. 64 light years from the Earth. The Vulpecula constellation is where this Exoplanet resides. A planet that rains glass. It has an orbital period of 2.2 days.

With each jump into hyperspace, Mase found himself further and further from the Earth. There were four more Exoplanets on his list.

WASP-76B

Unlike the other Exoplanets, Mase had to travel 634 light years to get here. Located in the Pisces Constellation. Its orbital period is 1.8 days. There was a debate that this planet rained iron.

KELT-9B

Through the vastness of space, Mase grew scared. He was all alone traveling further and further in the name of science. He traveled 670 light years to reach this next planet. The hottest exoplanet, its even hotter than some stars. Its Orbital Period is 1.5 days It is located in the Constellation of Cygnus.

TrES-2B

702 Light years away Mase found himself staring at the darkest planet ever discovered. Its orbital period is 2.5 days. Believed to be heated about 1800 degrees Fahrenheit. Located in the constellation Draco.

PSR B1620-26b

In the end, Mase managed to reach one of the oldest Exoplanets ever discovered. Located in the Scorpius constellation and about 12,390 light years from Earth. It is believed to be about 12.7 billion years old. It takes about 95 years to orbit its star.

As his journey concluded, Mase was excited to have learned all he could about Exoplanets. He had exhausted himself to the point of sleep. 'Til next time. Where will we find our little astronaut travelling to?

References:

55 Cancri E. (n.d.). NASA. https://science.nasa.gov/exoplanet-catalog/55-cancri-e/

GJ 1132 b – NASA Science. (n.d.). https://science.nasa.gov/exoplanet-catalog/gj-1132-b/

HD 189733 b – NASA Science. (n.d.). https://science.nasa.gov/exoplanet-catalog/hd-189733-b/

KELT-9 b – NASA Science. (n.d.). https://science.nasa.gov/exoplanet-catalog/kelt-9-b/

PSR B1620-26 b – NASA Science. (n.d.). https://science.nasa.gov/exoplanet-catalog/psr-b1620-26-b/

References:

TrES-2 b - NASA Science: (n.d.):
https://science.nasa.gov/exoplanet-catalog/tres-2-b/

Upsilon Andromedae b - NASA Science: (n.d.):
https://science.nasa.gov/exoplanet-catalog/upsilon-andromedae-b/

WASP-76 b - NASA Science: (n.d.):
https://science.nasa.gov/exoplanet-catalog/wasp-76-b/

www.ingramcontent.com/pod-product-compliance
Lightning Source LLC
Chambersburg PA
CBHW051950210526
45474CB00003B/78